T0199127

The King in Me

Isaiah Bealer

To order additional copies of this book, contact:
Xlibris
1-888-795-4274
www.Xlibris.com
Orders@Xlibris.com

Interior Image Credit: Brian Rivera

ISBN: 978-1-9845-8215-7 (sc)
ISBN: 978-1-9845-8214-0 (e)

Print information available on the last page

Rev. date: 08/05/2020

The King in Me

Written by Isaiah Bealer

"Daddy can you teach me my history?"

"Sure son, what exactly do you want to learn about our history?"

"I want to learn who are the great men in our history."

"I'm glad you want to learn that son. I will teach you all about them. Go in the closet and grab the gold crowns."

"Why did you want me to get the crowns dad?"

"Jeremiah, when we put these crowns on our head, they will allow us to travel back in time and meet great historical figures from the past."

"Wow, travel back in time?! Let's go!"

"Okay son, grab my hand. To travel safely we have to use all our focus and energy. Are you ready?"

"I'm ready!"

"Woah, dad where are we at?"

"We're in Egypt, which is a country in Africa. This country was originally called Kemet before it got the name Egypt. Do you see the pyramids over there?"

"Yes, I see the pyramids. Let's go talk to the man over there by the pyramids.

"Excuse me sir, who are you?"

"I am Ramses the Great. I am the Pharaoh over my country Kemet. The name Kemet means black earth."

"I want to be a ruler like you, can I?"

"Yes, you may but to be a true ruler, you must be sovereign. For this reason, my gift to you, is the jewel of sovereignty."

"Thank you, Pharaoh Ramses. I will cherish it forever!"

"You are welcome! Stay great, young royalty."

"I wonder where the crown bought us now dad."

"I don't know, but let's go talk to the man in front of all the gold and riches."

"Welcome to Mali. My name is Mansa Musa. I am the king in Mali."

"Wow, dad we met another ruler! Are you rich king Musa?"

"Yes, I am rich. It is said that I am the richest man to have ever lived. I also built schools, universities and libraries in Mali. I spread my wealth everywhere I go."

"Can I have some of your gold king Musa?"

"Yes, you may have some of my gold. I also will bless you with the jewel of prosperity. A king must first be prosperous in his mind to attain riches. Be great and travel safe my friends."

"Son it looks like the crowns bought us back to the states. From the look of the buildings, we are in the 1700's."

"Let's go talk to the man with the wooden clock in his hand dad."

"Hello, sir what is your name?"

"My name is Benjamin Banneker. Welcome to Baltimore, Maryland."

"Mr. Banneker what are you known for?"

"I am a mathematician and astronomer. I also was the first African American to write a Farmer's almanac. I am also the inventor of the clock here in America."

"Woah! You are super smart Mr. Banneker!"

"You are also, young man. To you, I give you the jewel of intelligence. To be successful in this life you must have spiritual and mental intelligence. Stay intelligent and be great my young friends."

"Thank you, Mr. Banneker!"

"Jeremiah now we've travelled to the windy city Chicago. Let's talk to the man by the sign."

"Welcome to the windy city. My name is Daniel Williams."

"Hello, Mr. Williams this is my son Jeremiah. My son and I are interested to know what in history you did that was great."

"I was the first person ever to perform a successful open-heart surgery. I did the operation right here in Chicago in 1893. I took pride doing the operation in this city because this city was founded by Jean Baptiste, who is a inspiration to me."

"That's amazing Mr. Williams! Were you nervous during the operation?"

"I was a little nervous, but my confidence overpowered it. To achieve your destiny, you must be confident. To you, I give you the jewel of confidence. Be confident and stay great!"

"I wonder why the crown bought us to a boxing ring dad?

"I believe we're about to meet a legendary boxer son!"

"Look behind you dad."

"Hello, my name is Jack Johnson. I am the first African American to win the World heavyweight title in 1908."

"That's cool! So, you used to knock people out Mr. Jack?"

"Yes, young man. I am the best boxer of my time. I also owned business and had patents on the tool known as the wrench."

"How did you do so much Mr. Jack?"

"I used the strength in my mind and built up an abundance of strength in my body. To you, young champ, I give you the jewel of strength. Be great and stay strong."

"Check out these trees son. We are on a tropical island of some sort."

"Welcome to Jamaica family. My name is Marcus Mosiah Garvey."

"It is an honor to meet you Mr. Garvey. Jeremiah this is one of the greatest leaders this world has ever seen."

"Woah, one of the greatest leaders ever! What did you achieve Mr. Garvey?"

"I was the first president of the organization the U.N.I.A. My organization encouraged Africans worldwide to unite and become entrepreneurs, inventors, teachers and builders of their own community."

"That's amazing Mr. Garvey! I'll be a great leader like you one day!"

"You already are! To help guide you along the way, I pass down the jewel of leadership! Be great and lead young king!"

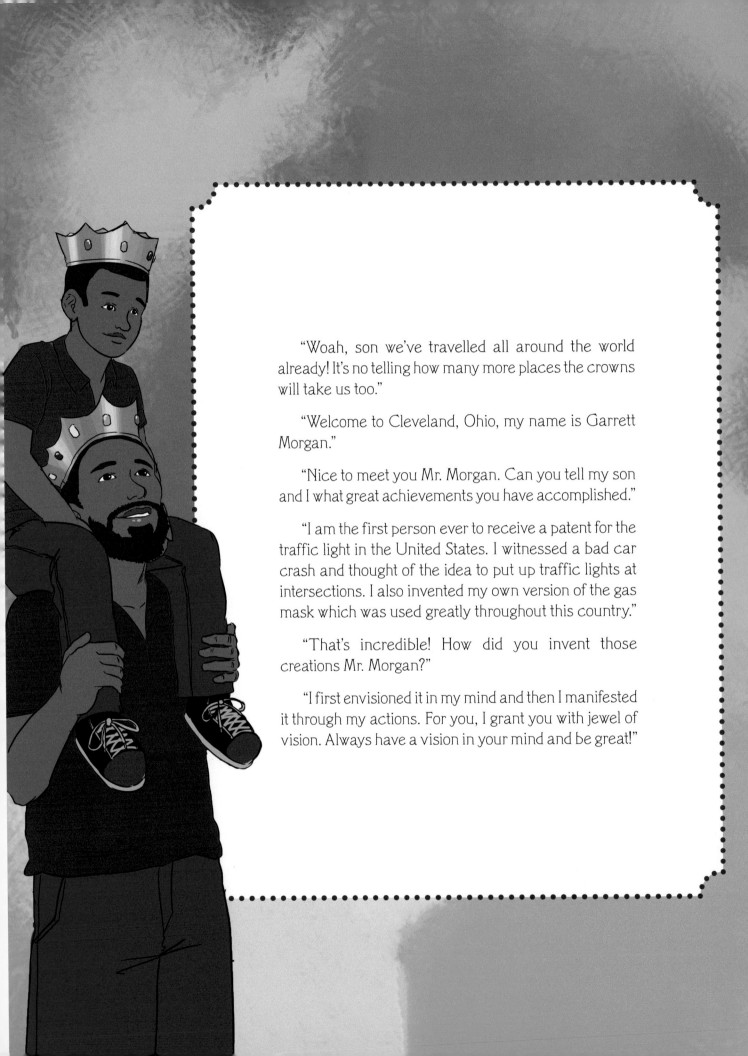

"Woah, son we've travelled all around the world already! It's no telling how many more places the crowns will take us too."

"Welcome to Cleveland, Ohio, my name is Garrett Morgan."

"Nice to meet you Mr. Morgan. Can you tell my son and I what great achievements you have accomplished."

"I am the first person ever to receive a patent for the traffic light in the United States. I witnessed a bad car crash and thought of the idea to put up traffic lights at intersections. I also invented my own version of the gas mask which was used greatly throughout this country."

"That's incredible! How did you invent those creations Mr. Morgan?"

"I first envisioned it in my mind and then I manifested it through my actions. For you, I grant you with jewel of vision. Always have a vision in your mind and be great!"

"Jeremiah, we've travelled to Los Angeles! We're in the Crenshaw district. Let's walk over in the plaza."

"Look dad, it's Nipsey Hussle!"

"Welcome to the Marathon store family! I'm honored that the crowns led you to me!"

"It's a pleasure for me and my son to meet you Nip!"

"Can I be a rapper like you Nip?"

"You can be whatever you want too young king. I studied the greats that came before me such as P.Diddy, Dame Dash, Master P, Ice Cube, Snoop and J Prince. Now artists like Meek Mill & I are taking action. Whatever you do in life, take ownership of it and always give back to your community. Crenshaw is where I grew up at and now, I own stores in this same neighborhood."

"I'll make sure I give back when I grow up! I promise!"

"That's what I like to hear. Life is a marathon Jeremiah, not a sprint. To help you along the way, I'm going to give you the jewel of longevity. Be great and run your marathon young king."

"Daddy, look now we're on the Lakers basketball court!"

"Yes, Jeremiah we are at the Staples Center! Look behind you!"

"Aww man! Look pops, it's Lebron and Kobe!"

"What's up! You want to play one on one?"

"I can't beat you, you're Kobe! You're the Black Mamba!

"Jeremiah, don't ever tell yourself what you can't do. Always tell yourself what you can do. Both Kobe and I whole life, we had to overcome obstacles and persevere. Now that we've overcome, we are playing for the same team Wilt, Kareem, Magic and Shaq played for. Some of the all- time greats!"

"I will make sure I do that! I will push through!"

"When hard times come, you must learn how to push through. In your crown, Lebron & I are placing the jewel of perseverance. Always persevere and be great!"

"Thanks Bron! Thanks Kobe!"

"Welcome to Oakland family!"

"Look, it's beast mode dad!"

"It's a pleasure for my son and I to meet you Marshawn. How are you doing bro?"

"I'm blessed fam. I can't complain about a thing."

"How do I feel the beast mode like you? I want to run even harder than you run."

"You don't feel beast mode, young king. Beast mode feels you. You activate the power from the inside of you and it transforms to the outside. This is how players like me, Deion Sanders, Bo Jackson and Kap were so dominant on the field. For your crown, I will give you the jewel of power. Be great and stay powerful."

"Thank you for the jewel Mr. Lynch."

"Woah, the crowns took us back to Jamaica again!"

"Sure did. Only this time son, I'm pretty sure were in modern times."

"Welcome to Jamaica mon!"

"Usain Bolt! You're the fastest man in the world!"

"Yes, for right now mon until you beat me in a race!"

"How'd you get so fast Mr. Bolt?"

"I was always naturally fast but my determination is what makes the difference."

"How do you get determination?

"Whatever you want bad enough, you will be determined to get it. To you, I pass the jewel of determination!"

"Thanks, Mr. Bolt!"

"These crowns are really some amazing travel guides son. After going travelling to Jamaica twice, now we've arrived in Atlanta."

"Hello family, I'm Jay Morrison. Some people refer to me as Mr. Real Estate."

"What's real estate Mr. Morrison?"

"Young king, real estate is physical property of land or buildings. The Atlanta Falcons football stadium behind me is considered real estate. I own real estate in numerous places. You also might know some southeren greats by the name of TIP, Micheal Render and David Banner who also buy property here in Georgia."

"Wow, so I can own the property that this stadium is built on?"

"Yes, you can own real estate in any part the world. The jewel I will bless you with is the jewel of ownership. Take ownership and be great young man."

"Are we in New York dad?"

"Yes son, I see a man standing in front of the sign that reads Queens, New York. Let's go talk to him."

"Hello, I'm Draymond John. I'm the starter of Fubu and several other businesses."

"It's a pleasure for me and my son to meet you Draymond."

"Will you invest in me Mr. Draymond?"

"I most certainly will Jeremiah, but the best investment you can receive comes from yourself. Legendary fashion moguls such as Dapper Dan had already begun to pave the way when I started my journey. When I started Fubu, I built it on faith and hard work. With that said, I invest into you the jewel of faith. Be great and walk in faith."

"Dad is this Times Square?"

"Yes son, we're in Times Square the heart of New York."

"Look dad, it's Jay-Z and Nas!"

"What's up young royalty, Jay and I are pleased to meet you!"

"Wow, you guys are the two of the greatest rappers ever!"

"Thank you for the compliment Jeremiah. Nas and I rapped along other greats such as Biggie, Pac, KRS One, the Lox and Rakim. Our skills helped us ascend to the top of the rap game but more importantly it was our knowledge."

"Jay is right. On behalf of me and Jay, we bless you with the jewel of knowledge. Stay sharp and stay great king."

"The crowns have bought us to a movie set. I wonder what actor we are about to meet."

"How are you gentleman doing this evening?

"We are great Mr. Washington and Mr. Jackson. It's an honor to meet the both of you!"

"Are you all about to shoot a movie?"

"Yes, we are young man. Combined the two of us have played over a hundred characters in our acting careers."

"Nevertheless, the most important characters we have ever played is ourselves. In this life, your character and actions represent who you are. To you young man, Denzell & I give you the jewel of character."

"Thank you, Mr. Jackson and Mr. Washington!"

"You are welcome young man. Samuel & I always want you to uphold your character and stay great!

"We made it back home safely son! What an adventurous journey!"

"I had fun dad and I learned so much about our history!"

"I'm glad you did son!"

"Dad, out of all the legends we met today, you are the one I look up to the most!"

"Thank you son. I always try to set the best example for you! Cherish the jewels in your crown and remember to always be great and embrace the king in you!"